OFFICIAL SQA PAST PAPERS WITH ANSWERS

STANDARD GRADE | CREDIT

MATHEMATICS
2008-2012

First exam published in 2008.
Published by Bright Red Publishing Ltd, 6 Stafford Street, Edinburgh EH3 7AU
tel: 0131 220 5804 fax: 0131 220 6710 info@brightredpublishing.co.uk www.brightredpublishing.co.uk

ISBN 978-1-84948-254-7

A CIP Catalogue record for this book is available from the British Library.

Bright Red Publishing is grateful to the copyright holders, as credited on the final page of the Question Section, for permission to use their material. Every effort has been made to trace the copyright holders and to obtain their permission for the use of copyright material. Bright Red Publishing will be happy to receive information allowing us to rectify any error or omission in future editions.

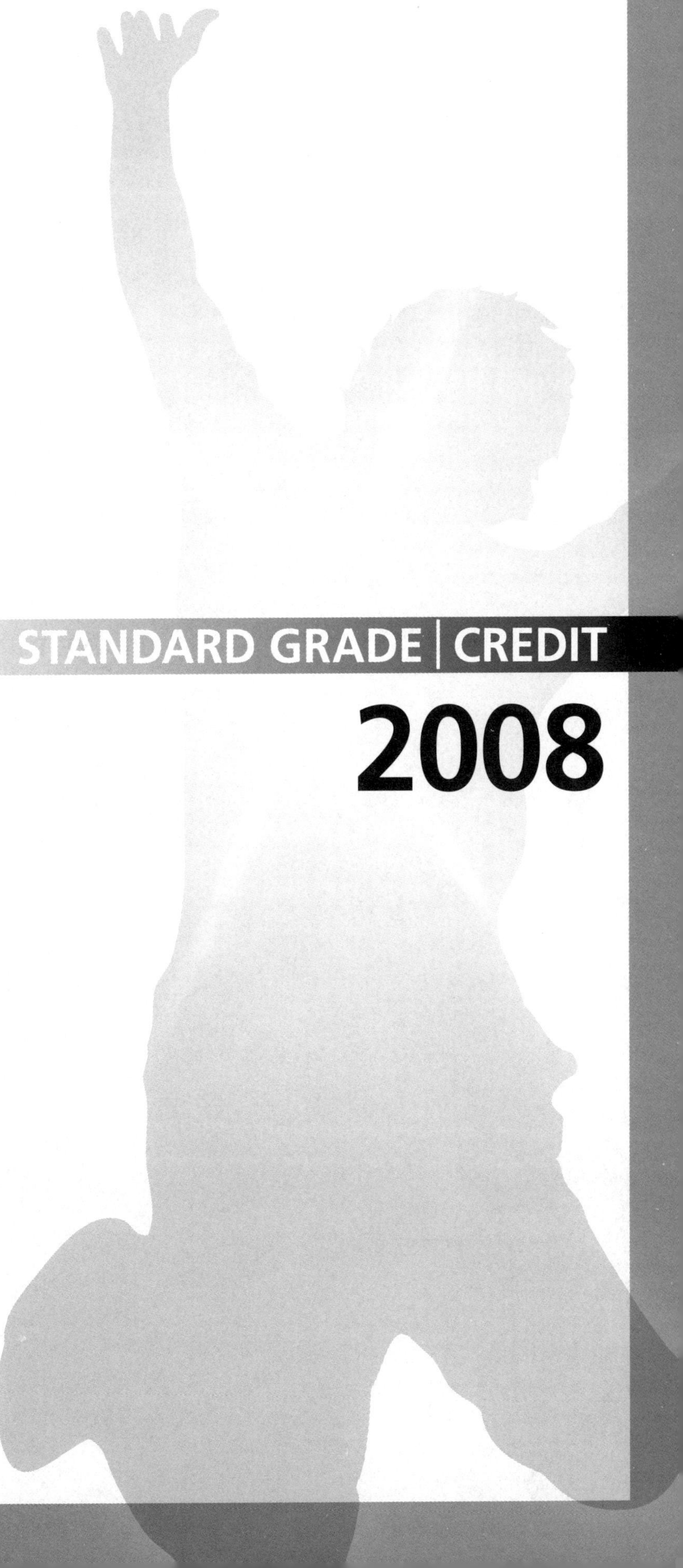

STANDARD GRADE | CREDIT

2008

[BLANK PAGE]

C

2500/405

NATIONAL
QUALIFICATIONS
2008

THURSDAY, 8 MAY
1.30 PM – 2.25 PM

MATHEMATICS
STANDARD GRADE
Credit Level
Paper 1
(Non-calculator)

1 **You may NOT use a calculator**.

2 Answer as many questions as you can.

3 Full credit will be given only where the solution contains appropriate working.

4 Square-ruled paper is provided.

FORMULAE LIST

The roots of $ax^2 + bx + c = 0$ are $x = \dfrac{-b \pm \sqrt{(b^2 - 4ac)}}{2a}$

Sine rule: $\dfrac{a}{\sin A} = \dfrac{b}{\sin B} = \dfrac{c}{\sin C}$

Cosine rule: $a^2 = b^2 + c^2 - 2bc \cos A$ or $\cos A = \dfrac{b^2 + c^2 - a^2}{2bc}$

Area of a triangle: Area $= \frac{1}{2}ab \sin C$

Standard deviation: $s = \sqrt{\dfrac{\sum(x - \bar{x})^2}{n-1}} = \sqrt{\dfrac{\sum x^2 - (\sum x)^2 / n}{n-1}}$, where n is the sample size.

KU | RE

1. Evaluate

$$24 \cdot 7 - 0 \cdot 63 \times 30.$$

2

2. Factorise fully

$$5x^2 - 45.$$

2

3. $$W = BH^2.$$

Change the subject of the formula to H.

2

4. A straight line cuts the x-axis at the point $(9, 0)$ and the y-axis at the point $(0, 18)$ as shown.

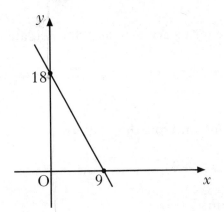

Find the equation of this line.

3

[Turn over

	KU	RE

5. Express as a single fraction in its simplest form

$$\frac{1}{p} + \frac{2}{(p+5)}.$$

2

6. Jane enters a two-part race.

(a) She cycles for 2 hours at a speed of $(x + 8)$ kilometres per hour.
Write down an expression in x for the distance cycled.

1

(b) She then runs for 30 minutes at a speed of x kilometres per hour.
Write down an expression in x for the distance run.

1

(c) The **total** distance of the race is 46 kilometres.
Calculate Jane's **running** speed.

3

7. The 4th term of each number pattern below is the **mean** of the previous three terms.

(a) When the first three terms are 1, 6, and 8, calculate the 4th term.

1

(b) When the first three terms are x, $(x + 7)$ and $(x + 11)$, calculate the 4th term.

1

(c) When the first, second and fourth terms are

$$-2x, \qquad (x + 5), \qquad \underline{\hspace{1cm}}, \qquad (2x + 4),$$

calculate the 3rd term.

2

KU | RE

8. The curved part of the letter A in the *Artwork* logo is in the shape of a parabola.

The equation of this parabola is $y = (x - 8)(2 - x)$.

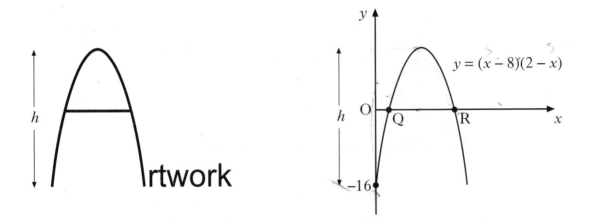

(a) Write down the coordinates of Q and R.

2

(b) Calculate the height, h, of the letter A.

3

9. Simplify

$$m^3 \times \sqrt{m}.$$

2

[Turn over

10. Part of the graph of $y = a^x$, where $a > 0$, is shown below.

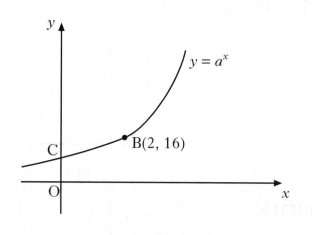

The graph cuts the y-axis at C.

(*a*) Write down the coordinates of C.

1

B is the point (2, 16).

(*b*) Calculate the value of a.

2

11. A right angled triangle has dimensions as shown.

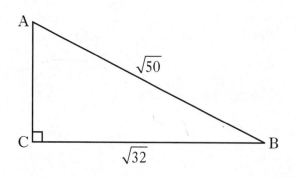

Calculate the length of AC, leaving your answer as a surd **in its simplest form**.

3

	KU	RE

12. Given that

$$x^2 - 10x + 18 = (x - a)^2 + b,$$

find the values of a and b.

3

13. A new fraction is obtained by adding x to the numerator and denominator of the fraction $\frac{17}{24}$.

This new fraction is equivalent to $\frac{2}{3}$.

Calculate the value of x.

3

[END OF QUESTION PAPER]

[BLANK PAGE]

C

2500/406

NATIONAL
QUALIFICATIONS
2008

THURSDAY, 8 MAY
2.45 PM – 4.05 PM

MATHEMATICS
STANDARD GRADE
Credit Level
Paper 2

1 **You may use a calculator**.

2 Answer as many questions as you can.

3 Full credit will be given only where the solution contains appropriate working.

4 Square-ruled paper is provided.

FORMULAE LIST

The roots of $ax^2 + bx + c = 0$ are $x = \dfrac{-b \pm \sqrt{(b^2 - 4ac)}}{2a}$

Sine rule: $\dfrac{a}{\sin A} = \dfrac{b}{\sin B} = \dfrac{c}{\sin C}$

Cosine rule: $a^2 = b^2 + c^2 - 2bc \cos A$ or $\cos A = \dfrac{b^2 + c^2 - a^2}{2bc}$

Area of a triangle: Area $= \frac{1}{2} ab \sin C$

Standard deviation: $s = \sqrt{\dfrac{\sum (x - \bar{x})^2}{n - 1}} = \sqrt{\dfrac{\sum x^2 - (\sum x)^2 / n}{n - 1}}$, where n is the sample size.

	KU	RE

1. A local council recycles 42 000 tonnes of waste a year.

    The council aims to increase the amount of waste recycled by 8% each year.

    How much waste does it expect to recycle in 3 years time?

    Give your answer **to three significant figures**.   4

2. In a class, 30 pupils sat a test.

    The marks are illustrated by the stem and leaf diagram below.

    Test Marks

    ```
    0 | 9
    1 | 6 6 7 8
    2 | 0 4 5 7 9 9 9
    3 | 2 2 3 5 5 6 8
    4 | 0 2 3 4 5 5 7 7 8
    5 | 0 0
    ```

    n = 30   1 | 6 = 16

    (a) Write down the median and the modal mark.   2

    (b) Find the probability that a pupil selected at random scored **at least** 40 marks.   1

3. In a sale, all cameras are reduced by 20%.

    A camera now costs £45.

    Calculate the **original** cost of the camera.   3

    NOW £45

[Turn over

KU | RE

4. Aaron saves 50 pence and 20 pence coins in his piggy bank.

Let x be the number of 50 pence coins in his bank.

Let y be the number of 20 pence coins in his bank.

(a) There are 60 coins in his bank.

Write down an equation in x and y to illustrate this information.

1

(b) The total value of the coins is £17·40.

Write down another equation in x and y to illustrate this information.

1

(c) Hence find **algebraically** the number of 50 pence coins Aaron has in his piggy bank.

3

5. A circle, centre the origin, is shown.

P is the point (8, 1).

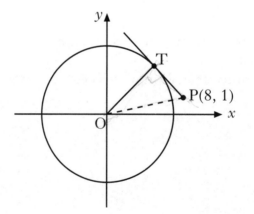

(a) Calculate the length of OP.

2

The diagram also shows a tangent from P which touches the circle at T.
The radius of the circle is 5 units.

(b) Calculate the length of PT.

2

KU | RE

6. The distance, *d* kilometres, to the horizon, when viewed from a cliff top, varies directly as the square root of the height, *h* metres, of the cliff top above sea level.

From a cliff top 16 metres above sea level, the distance to the horizon is 14 kilometres.

A boat is 20 kilometres from a cliff whose top is 40 metres above sea level.

Is the boat beyond the horizon?

Justify your answer.

5

7. A telegraph pole is 6·2 metres high.

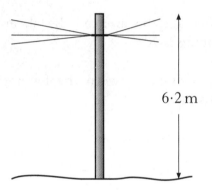

6·2 m

The wind blows the pole over into the position as shown below.

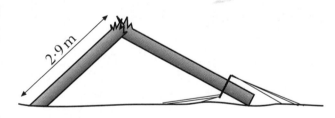

2·9 m

2·9 m

B

130°

A

C

AB is 2·9 metres and angle ABC is 130°.

Calculate the length of AC.

4

[Turn over

KU | RE

8. A farmer builds a sheep-pen using two lengths of fencing and a wall.

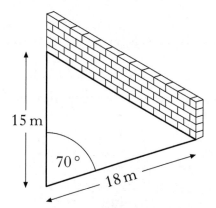

The two lengths of fencing are 15 metres and 18 metres long.

(a) Calculate the area of the sheep-pen, when the angle between the fencing is 70°.

3

(b) What angle between the fencing would give the farmer the largest possible area?

1

9. Contestants in a quiz have 25 seconds to answer a question.

This time is indicated on the clock.

The tip of the clock hand moves through the arc AB as shown.

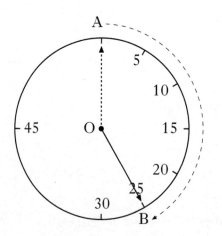

(a) Calculate the size of angle AOB.

1

(b) The length of arc AB is 120 centimetres.

Calculate the length of the clock hand.

4

KU | RE

10. To hire a car costs £25 per day plus a mileage charge.

The first 200 miles are free with each additional mile charged at 12 pence.

CAR HIRE

£25 per day

• **first 200** miles free
• each additional mile only 12p

(a) Calculate the cost of hiring a car for 4 days when the mileage is 640 miles.

1

(b) A car is hired for d days and the mileage is m miles where $m > 200$.

Write down a formula for the cost £C of hiring the car.

3

11. The minimum number of roads joining 4 towns to each other is 6 as shown.

The minimum number of roads, r, joining n towns to each other is given by the formula
$$r = \tfrac{1}{2}n(n-1).$$

(a) State the minimum number of roads needed to join 7 towns to each other.

1

(b) When $r = 55$, show that $n^2 - n - 110 = 0$.

2

(c) Hence find **algebraically** the value of n.

3

[Turn over for Question 12 on *Page eight*

KU | RE

12. The diagram shows part of the graph of $y = \tan x°$.

The line $y = 5$ is drawn and intersects the graph of $y = \tan x°$ at P and Q.

(a) Find the x-coordinates of P and Q.

3

(b) Write down the x-coordinate of the point R, where the line $y = 5$ next intersects the graph of $y = \tan x°$.

1

[END OF QUESTION PAPER]

[BLANK PAGE]

C

2500/405

NATIONAL
QUALIFICATIONS
2009

WEDNESDAY, 6 MAY
1.30 PM – 2.25 PM

MATHEMATICS
STANDARD GRADE
Credit Level
Paper 1
(Non-calculator)

1 **You may NOT use a calculator**.

2 Answer as many questions as you can.

3 Full credit will be given only where the solution contains appropriate working.

4 Square-ruled paper is provided.

FORMULAE LIST

The roots of $ax^2 + bx + c = 0$ are $x = \dfrac{-b \pm \sqrt{(b^2 - 4ac)}}{2a}$

Sine rule: $\dfrac{a}{\sin A} = \dfrac{b}{\sin B} = \dfrac{c}{\sin C}$

Cosine rule: $a^2 = b^2 + c^2 - 2bc \cos A$ or $\cos A = \dfrac{b^2 + c^2 - a^2}{2bc}$

Area of a triangle: Area $= \frac{1}{2} ab \sin C$

Standard deviation: $s = \sqrt{\dfrac{\sum(x - \bar{x})^2}{n-1}} = \sqrt{\dfrac{\sum x^2 - (\sum x)^2 / n}{n-1}}$, where n is the sample size.

	KU	RE

1. Evaluate

$$(846 \div 30) - 1 \cdot 09.$$

KU 2

2. Evaluate

$$4\frac{1}{3} - 1\frac{1}{2}.$$

KU 2

3. Given that

$$f(x) = x^2 + 3,$$

(a) evaluate $f(-4)$

KU 2

(b) find t when $f(t) = 52$.

RE 2

4. (a) Factorise

$$x^2 - 4y^2.$$

KU 1

(b) Expand and simplify

$$(2x - 1)(x + 4).$$

KU 1

(c) Expand

$$x^{\frac{1}{2}}\left(3x + x^{-2}\right).$$

KU 2

[Turn over

KU | RE

5. In triangle ABC:

- angle ACB = 90°
- AB = 8 centimetres
- AC = 4 centimetres.

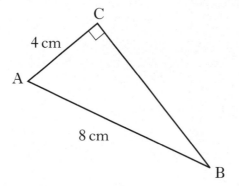

Calculate the length of BC.

Give your answer **as a surd in its simplest form**.

3

6. There are 4 girls and 14 boys in a class.

A child is chosen at random and is asked to roll a die, numbered 1 to 6.

Which of these is more likely?

  A: the child is female.

   OR

  B: the child rolls a 5.

Justify your answer.

3

7. This year, Ben paid £260 for his car insurance.

This is an increase of 30% on last year's payment.

How much did Ben pay last year?

3

KU RE

8. In triangle PQR:

   • PQ = x centimetres

   • PR = $5x$ centimetres

   • QR = $2y$ centimetres.

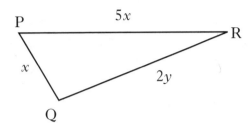

(a) The perimeter of the triangle is 42 centimetres.

   Write down an equation in x and y to illustrate this information.

   2

(b) PR is 2 centimetres longer than QR.

   Write down another equation in x and y to illustrate this information.

   2

(c) Hence calculate the values of x and y.

   3

9. A formula used to calculate the flow of water in a pipe is

$$f = \frac{kd^2}{20}.$$

   Change the subject of the formula to d.

   3

[Turn over

KU | RE

10. The diagram below shows the path of a rocket which is fired into the air.

The height, h metres, of the rocket after t seconds is given by

$$h(t) = -2t(t - 14).$$

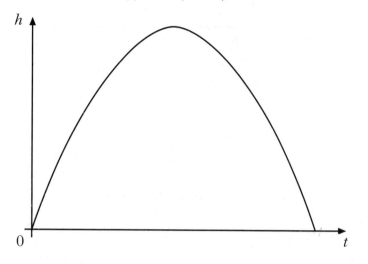

(a) For how many seconds is the rocket in flight? 2

(b) What is the maximum height reached by the rocket? 2

KU | RE

11. In triangle ABC:

- BC = 6 metres
- AC = 10 metres
- angle ABC = 30°.

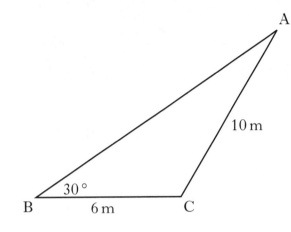

Given that sin 30° = 0·5, show that sin A = 0·3.

3

[END OF QUESTION PAPER]

[BLANK PAGE]

C

2500/406

NATIONAL
QUALIFICATIONS
2009

WEDNESDAY, 6 MAY
2.45 PM – 4.05 PM

MATHEMATICS
STANDARD GRADE
Credit Level
Paper 2

1 **You may use a calculator**.

2 Answer as many questions as you can.

3 Full credit will be given only where the solution contains appropriate working.

4 Square-ruled paper is provided.

FORMULAE LIST

The roots of $ax^2 + bx + c = 0$ are $x = \dfrac{-b \pm \sqrt{(b^2 - 4ac)}}{2a}$

Sine rule: $\dfrac{a}{\sin A} = \dfrac{b}{\sin B} = \dfrac{c}{\sin C}$

Cosine rule: $a^2 = b^2 + c^2 - 2bc \cos A$ or $\cos A = \dfrac{b^2 + c^2 - a^2}{2bc}$

Area of a triangle: Area $= \frac{1}{2}ab \sin C$

Standard deviation: $s = \sqrt{\dfrac{\sum(x - \bar{x})^2}{n-1}} = \sqrt{\dfrac{\sum x^2 - (\sum x)^2 / n}{n-1}}$, where n is the sample size.

KU | RE

1. One atom of gold weighs $3 \cdot 27 \times 10^{-22}$ grams.

   How many atoms will there be in one kilogram of gold?

   Give your answer **in scientific notation correct to 2 significant figures**.

   3

2. Lemonade is to be poured from a 2 litre bottle into glasses.

   Each glass is in the shape of a cylinder of radius 3 centimetres and height 8 centimetres.

   How many full glasses can be poured from the bottle?

   4

3. Solve the quadratic equation $x^2 - 4x - 6 = 0$.

   Give your answers **correct to 1 decimal place**.

   4

[Turn over

KU | RE

4. Two fridge magnets are mathematically similar.

One magnet is 4 centimetres long and the other is 10 centimetres long.

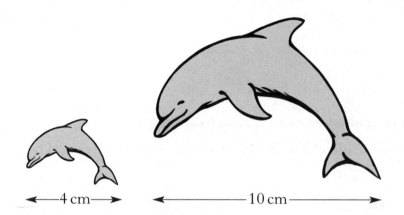

←—4 cm—→   ←————10 cm————→

The area of the smaller magnet is 18 square centimetres.

Calculate the area of the larger magnet.

3

5. Tom looked at the cost of 10 different flights to New York.

He calculated that the mean cost was £360 and the standard deviation was £74.

A tax of £12 is then added to each flight.

Write down the new mean and standard deviation.

2

KU | RE

6. Teams in a quiz answer questions on film and sport.

This scatter graph shows the scores of some of the teams.

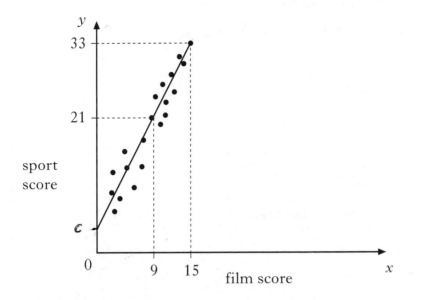

A line of best fit is drawn as shown above.

(a) Find the equation of this straight line.   4

(b) Use this equation to estimate the sport score for a team with a film score of 20.   2

7. (a) The air temperature, $t°$ Celsius, varies inversely as the square of the distance, d metres, from a furnace.

Write down a formula connecting t and d.   2

(b) At a distance of 2 metres from the furnace, the air temperature is 50 °C.

Calculate the air temperature at a distance of 5 metres from the furnace.   3

[Turn over

KU | RE

8. A company makes large bags of crisps which contain 90 grams of fat.

   The company aims to reduce the fat content of the crisps by 50%.

   They decide to reduce the fat content by 20% each year.

   Will they have achieved their aim by the end of the 3rd year?

   Justify your answer.

   4

9. Jane is taking part in an orienteering competition.

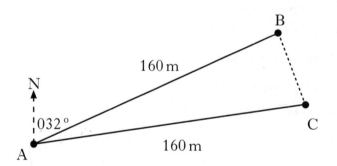

   She should have run 160 metres from A to B on a bearing of 032°.

   However, she actually ran 160 metres from A to C on a bearing of 052°.

   (a) Write down the size of angle BAC.

   1

   (b) Calculate the length of BC.

   3

   (c) What is the bearing from C to B?

   2

10. The weight, W kilograms, of a giraffe is related to its age, M months, by the formula

$$W = \tfrac{1}{4}\left(M^2 - 4M + 272\right).$$

At what age will a giraffe weigh 83 kilograms?

4

11. A cone is formed from a paper circle with a sector removed as shown.

The radius of the paper circle is 30 cm.

Angle AOB is 100°.

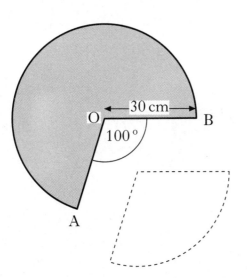

(a) Calculate the area of paper used to make the cone.

3

(b) Calculate the circumference of the base of the cone.

3

[Turn over for Question 12 on *Page eight*

KU | RE

12. The nth term, T_n of the sequence 1, 3, 6, 10, . . . is given by the formula:

$$T_n = \frac{1}{2}n(n+1)$$

1st term $T_1 = \frac{1}{2} \times 1(1+1) = 1$

2nd term $T_2 = \frac{1}{2} \times 2(2+1) = 3$

3rd term $T_3 = \frac{1}{2} \times 3(3+1) = 6$

(a) Calculate the 20th term, T_{20}.

1

(b) Show that $T_{n+1} = \frac{1}{2}\left(n^2 + 3n + 2\right)$.

2

(c) Show that $T_n + T_{n+1}$ is a square number.

2

[END OF QUESTION PAPER]

[BLANK PAGE]

C

2500/405

NATIONAL QUALIFICATIONS 2010

WEDNESDAY, 5 MAY 1.30 PM – 2.25 PM

MATHEMATICS STANDARD GRADE Credit Level Paper 1 (Non-calculator)

1 **You may NOT use a calculator**.

2 Answer as many questions as you can.

3 Full credit will be given only where the solution contains appropriate working.

4 Square-ruled paper is provided.

FORMULAE LIST

The roots of $ax^2 + bx + c = 0$ are $x = \dfrac{-b \pm \sqrt{(b^2 - 4ac)}}{2a}$

Sine rule: $\dfrac{a}{\sin A} = \dfrac{b}{\sin B} = \dfrac{c}{\sin C}$

Cosine rule: $a^2 = b^2 + c^2 - 2bc \cos A$ or $\cos A = \dfrac{b^2 + c^2 - a^2}{2bc}$

Area of a triangle: Area $= \frac{1}{2} ab \sin C$

Standard deviation: $s = \sqrt{\dfrac{\sum (x - \bar{x})^2}{n - 1}} = \sqrt{\dfrac{\sum x^2 - (\sum x)^2 / n}{n - 1}}$, where n is the sample size.

	KU	RE

1. Evaluate

$$40\% \text{ of } £11 \cdot 50 - £1 \cdot 81.$$

KU: 2

2. Evaluate

$$\frac{2}{5} \div 1\frac{1}{10}.$$

KU: 2

3. Change the subject of the formula to s.

$$t = \frac{7s + 4}{2}.$$

KU: 3

4. Two functions are given below.

$$f(x) = x^2 - 4x$$

$$g(x) = 2x + 7$$

(a) If $f(x) = g(x)$, show that $x^2 - 6x - 7 = 0$.

RE: 2

(b) Hence find **algebraically** the values of x for which $f(x) = g(x)$.

RE: 2

[Turn over

KU | RE

5. A bag contains 27 marbles. Some are black and some are white.

The probability that a marble chosen at random is black is $\frac{4}{9}$.

(a) What is the probability that a marble chosen at random is white?   1

(b) How many white marbles are in the bag?   1

6. Cleano washing powder is on special offer.

Each box on special offer contains 20% more powder than the standard box.

A box on special offer contains 900 grams of powder.

How many grams of powder does the standard box contain?   3

	KU	RE

7. A straight line has equation $y = mx + c$, where m and c are constants.

(a) The point (2, 7) lies on this line.

Write down an equation in m and c to illustrate this information.   **1**

(b) A second point (4, 17) also lies on this line.

Write down another equation in m and c to illustrate this information.   **1**

(c) Hence calculate the values of m and c.   **3**

(d) Write down the gradient of this line.   **1**

8. (a) Simplify $\sqrt{2} \times \sqrt{18}$.   **1**

(b) Simplify $\sqrt{2} + \sqrt{18}$.   **1**

(c) Hence show that $\dfrac{\sqrt{2} \times \sqrt{18}}{\sqrt{2} + \sqrt{18}} = \dfrac{3\sqrt{2}}{4}$.   **2**

[Turn over

KU RE

9. Part of the graph of the straight line with equation $y = \frac{1}{3}x + 2$, is shown below.

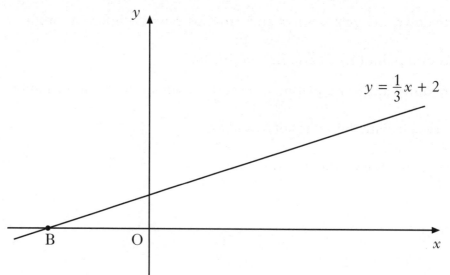

(a) Find the coordinates of the point B.

2

(b) For what values of x is $y < 0$?

1

10. A number pattern is shown below.

$$1^3 = \frac{1^2 \times 2^2}{4}$$

$$1^3 + 2^3 = \frac{2^2 \times 3^2}{4}$$

$$1^3 + 2^3 + 3^3 = \frac{3^2 \times 4^2}{4}$$

(a) Write down a similar expression for $1^3 + 2^3 + 3^3 + 4^3 + 5^3$.

(b) Write down a similar expression for $1^3 + 2^3 + 3^3 + \ldots + n^3$.

(c) Hence **evaluate** $1^3 + 2^3 + 3^3 + \ldots + 9^3$.

11. Two triangles have dimensions as shown.

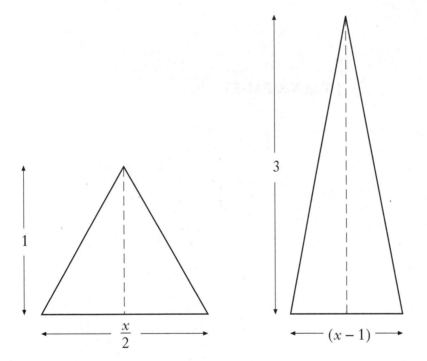

The triangles are equal in area.

Calculate the value of x.

[END OF QUESTION PAPER]

[BLANK PAGE]

C

2500/406

| NATIONAL QUALIFICATIONS 2010 | WEDNESDAY, 5 MAY 2.45 PM – 4.05 PM | MATHEMATICS STANDARD GRADE Credit Level Paper 2 |

1 **You may use a calculator**.

2 Answer as many questions as you can.

3 Full credit will be given only where the solution contains appropriate working.

4 Square-ruled paper is provided.

FORMULAE LIST

The roots of $ax^2 + bx + c = 0$ are $x = \dfrac{-b \pm \sqrt{(b^2 - 4ac)}}{2a}$

Sine rule: $\dfrac{a}{\sin A} = \dfrac{b}{\sin B} = \dfrac{c}{\sin C}$

Cosine rule: $a^2 = b^2 + c^2 - 2bc \cos A$ or $\cos A = \dfrac{b^2 + c^2 - a^2}{2bc}$

Area of a triangle: Area $= \frac{1}{2}ab \sin C$

Standard deviation: $s = \sqrt{\dfrac{\sum(x - \bar{x})^2}{n-1}} = \sqrt{\dfrac{\sum x^2 - (\sum x)^2 / n}{n-1}}$, where n is the sample size.

	KU	RE

1. It is estimated that an iceberg weighs 84 000 tonnes.

   As the iceberg moves into warmer water, its weight decreases by 25% each day.

   What will the iceberg weigh after 3 days in the warmer water?

   Give your answer **correct to three significant figures**.

   KU: 4

2. Expand fully and simplify

$$x(x-1)^2.$$

   KU: 2

3. A machine is used to put drawing pins into boxes.

   A sample of 8 boxes is taken and the number of drawing pins in each is counted.

   The results are shown below:

102	102	101	98	99	101	103	102

   (a) Calculate the mean and standard deviation of this sample.

   KU: 3

   (b) A sample of 8 boxes is taken from another machine.

   This sample has a mean of 103 and a standard deviation of 2·1.

   Write down two valid comparisons between the samples.

   RE: 2

4. Use the quadratic formula to solve the equation,

$$3x^2 + 5x - 7 = 0.$$

   Give your answers correct to **1 decimal place**.

   KU: 4

[Turn over

KU | RE

5. A concrete ramp is to be built.

The ramp is in the shape of a cuboid and a triangular prism with dimensions as shown.

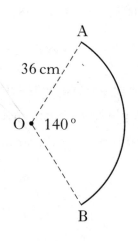

0·5 m

1 m

2 m

x

2 m

(a) Calculate the value of x.   2

(b) Calculate the volume of concrete required to build the ramp.   3

6. A circle, centre O, has radius 36 centimetres.

Part of this circle is shown.

Angle AOB = 140°.

A

36 cm

O   140°

B

Calculate the length of arc AB.   3

7. Shampoo is available in travel size and salon size bottles.

The bottles are mathematically similar.

travel   salon

The travel size contains 200 millilitres and is 12 centimetres in height.

The salon size contains 1600 millilitres.

Calculate the height of the salon size bottle.

3

[Turn over

KU | RE

8. As part of their training, footballers run around a triangular circuit DEF.

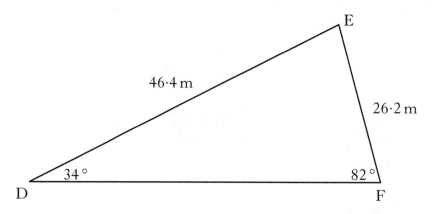

- ∠EDF = 34°

- ∠DFE = 82°

- DE = 46·4 metres

- EF = 26·2 metres

How many **complete** circuits must they run to cover **at least** 1000 metres?

4

9. The ratio of sugar to fruit in a particular jam is 5 : 4.

It is decided to:

- **decrease** the sugar content by 20%

- **increase** the fruit content by 20%.

Calculate the new ratio of sugar to fruit.

Give your answer in its simplest form.

4

KU | RE

10. In triangle PQR:

  • PQ = 5 centimetres

  • PR = 6 centimetres

  • area of triangle PQR = 12 square centimetres

  • angle QPR is **obtuse**.

Calculate the size of angle QPR.

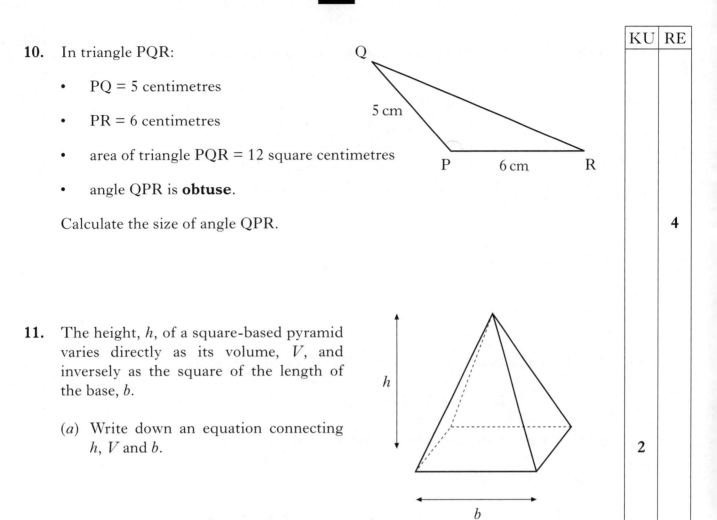

4

11. The height, h, of a square-based pyramid varies directly as its volume, V, and inversely as the square of the length of the base, b.

(a) Write down an equation connecting h, V and b.

2

A square-based pyramid of height 12 centimetres has a volume of 256 cubic centimetres and length of base 8 centimetres.

(b) Calculate the height of a square-based pyramid of volume 600 cubic centimetres and length of base 10 centimetres.

3

[Turn over for Questions 12 and 13 on *Page eight*

KU | RE

12. A right-angled triangle has dimensions, in centimetres, as shown.

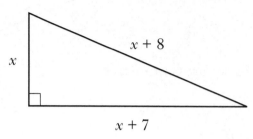

Calculate the value of x.

5

13. The depth of water, D metres, in a harbour is given by the formula

$$D = 3 + 1 \cdot 75 \sin 30 h^\circ$$

where h is the number of hours after midnight.

(a) Calculate the depth of water at 5 am.

2

(b) Calculate the maximum difference in depth of the water in the harbour.

Do not use a trial and improvement method.

2

[END OF QUESTION PAPER]

[BLANK PAGE]

C

2500/405

NATIONAL	WEDNESDAY, 4 MAY	MATHEMATICS
QUALIFICATIONS	1.30 PM – 2.25 PM	STANDARD GRADE
2011		Credit Level
		Paper 1
		(Non-calculator)

1 **You may NOT use a calculator**.

2 Answer as many questions as you can.

3 Full credit will be given only where the solution contains appropriate working.

4 Square-ruled paper is provided. If you make use of this, you should write your name on it clearly and put it inside your answer booklet.

FORMULAE LIST

The roots of $ax^2 + bx + c = 0$ are $x = \dfrac{-b \pm \sqrt{(b^2 - 4ac)}}{2a}$

Sine rule: $\dfrac{a}{\sin A} = \dfrac{b}{\sin B} = \dfrac{c}{\sin C}$

Cosine rule: $a^2 = b^2 + c^2 - 2bc \cos A$ or $\cos A = \dfrac{b^2 + c^2 - a^2}{2bc}$

Area of a triangle: Area $= \frac{1}{2}ab \sin C$

Standard deviation: $s = \sqrt{\dfrac{\sum (x - \bar{x})^2}{n - 1}} = \sqrt{\dfrac{\sum x^2 - (\sum x)^2 / n}{n - 1}}$, where n is the sample size.

	KU	RE

1. Evaluate

$$2 \cdot 4 + 5 \cdot 46 \div 60.$$

(KU: 2)

2. Factorise fully

$$2m^2 - 18.$$

(KU: 2)

3. Given that $f(x) = 5 - x^2$, evaluate $f(-3)$.

(KU: 2)

4. Solve the equation

$$3x + 1 = \frac{x-5}{2}.$$

(KU: 3)

[Turn over

5. Jamie is going to bake cakes for a party.

He needs $\frac{2}{5}$ of a block of butter for 1 cake.

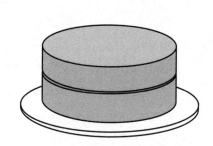

He has 7 blocks of butter.

How many cakes can Jamie bake?

3

6. A driving examiner looks at her diary for the next 30 days.

She writes down the number of driving tests booked for each day as shown below.

Number of tests booked	0	1	2	3	4	5	6
Frequency	1	1	3	2	9	10	4

(a) Find the median for this data.

2

(b) Find the probability that **more than** 4 tests are booked for one day.

1

KU | RE

7. (*a*) Brian, Molly and their four children visit Waterworld.

The total cost of their tickets is £56.

Let *a* pounds be the cost of an adult's ticket and *c* pounds the cost of a child's ticket.

Write down an equation in terms of *a* and *c* to illustrate this information.

1

(*b*) Sarah and her three children visit Waterworld.

The total cost of their tickets is £36.

Write down another equation in terms of *a* and *c* to illustrate this information.

1

(*c*) (i) Calculate the cost of a child's ticket.

2

(ii) Calculate the cost of an adult's ticket.

1

[Turn over

KU | RE

8. A square, OSQR, is shown below.

Q is the point (8, 8).

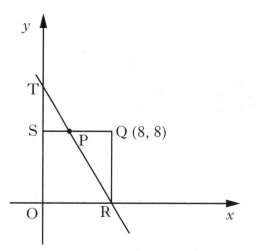

The straight line TR cuts the y-axis at T (0, 12) and the x-axis at R.

(*a*) Find the equation of the line TR.

3

The line TR also cuts SQ at P.

(*b*) Find the coordinates of P.

4

9. (*a*) Simplify $2a \times a^{-4}$.

1

(*b*) Solve for x, $\sqrt{x} + \sqrt{18} = 4\sqrt{2}$.

3

KU RE

10. In triangle ABC

- AC = 4 centimetres
- BC = 10 centimetres
- angle BAC = 150°

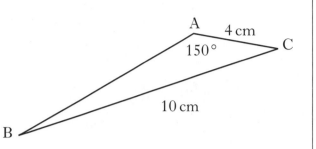

Given that $\sin 30° = \dfrac{1}{2}$, show that $\sin B = \dfrac{1}{5}$.

4

11. F varies directly as s and inversely as the square of d.

(a) Write down a relationship connecting F, s and d.

1

(b) What is the effect on F when s is halved and d is doubled?

3

12. The sums, S_2, S_3 and S_4 of the first 2, 3 and 4 natural numbers are given by:

$$S_2 = 1 + 2 \qquad = \frac{1}{2}\,(2 \times 3) = 3$$

$$S_3 = 1 + 2 + 3 \qquad = \frac{1}{2}\,(3 \times 4) = 6$$

$$S_4 = 1 + 2 + 3 + 4 \quad = \frac{1}{2}\,(4 \times 5) = 10$$

(a) Find S_{10}, the sum of the first 10 natural numbers.

1

(b) Write down the formula for the sum, S_n, of the first n natural numbers.

1

[END OF QUESTION PAPER]

[BLANK PAGE]

C

2500/406

| NATIONAL QUALIFICATIONS 2011 | WEDNESDAY, 4 MAY 2.45 PM – 4.05 PM | MATHEMATICS STANDARD GRADE Credit Level Paper 2 |

1 **You may use a calculator**.

2 Answer as many questions as you can.

3 Full credit will be given only where the solution contains appropriate working.

4 Square-ruled paper is provided. If you make use of this, you should write your name on it clearly and put it inside your answer booklet.

FORMULAE LIST

The roots of $ax^2 + bx + c = 0$ are $x = \dfrac{-b \pm \sqrt{(b^2 - 4ac)}}{2a}$

Sine rule: $\dfrac{a}{\sin A} = \dfrac{b}{\sin B} = \dfrac{c}{\sin C}$

Cosine rule: $a^2 = b^2 + c^2 - 2bc \cos A$ or $\cos A = \dfrac{b^2 + c^2 - a^2}{2bc}$

Area of a triangle: Area $= \frac{1}{2}ab \sin C$

Standard deviation: $s = \sqrt{\dfrac{\sum(x - \bar{x})^2}{n - 1}} = \sqrt{\dfrac{\sum x^2 - (\sum x)^2 / n}{n - 1}}$, where n is the sample size.

1. Olga normally runs a total distance of 28 miles per week.

    She decides to increase her distance by 10% a week for the next four weeks.

    How many miles will she run in the fourth week?

    3

2. Expand and simplify

    $$(3x + 1)(x^2 - 5x + 4).$$

    3

3. Solve the equation

    $$2x^2 + 3x - 7 = 0.$$

    Give your answers **correct to 2 significant figures**.

    4

4. A car is valued at £3780.

    This is 16% less than last year's value.

    What was the value of the car last year?

    3

	KU	RE

[Turn over

KU | RE

5. A spiral staircase is being designed.

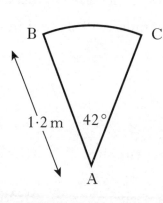

Each step is made from a sector of a circle as shown.

The radius is 1·2 metres.

Angle BAC is 42°.

For the staircase to pass safety regulations, the arc BC must be at least 0·9 metres.

Will the staircase pass safety regulations?

4

6. Two rectangular solar panels, A and B, are mathematically similar.

Panel A has a diagonal of 90 centimetres and an area of 4020 square centimetres.

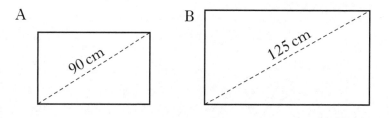

A salesman claims that panel B, with a diagonal of 125 centimetres, will be double the area of panel A.

Is this claim justified?

Show all your working.

4

KU | RE

7. ABCDE is a regular pentagon with each side 1 centimetre.

Angle CDF is 72°.

EDF is a straight line.

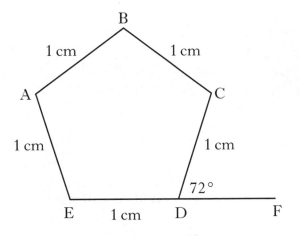

(a) Write down the size of angle ABC.

1

(b) Calculate the length of AC.

3

8. A pipe has water in it as shown.

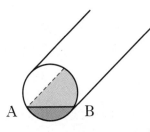

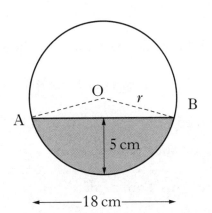

The depth of the water is 5 centimetres.

The width of the water surface, AB, is 18 centimetres.

Calculate r, the radius of the pipe.

3

[Turn over

KU | RE

9. A flower planter is in the shape of a prism.

The cross-section is a trapezium with dimensions as shown.

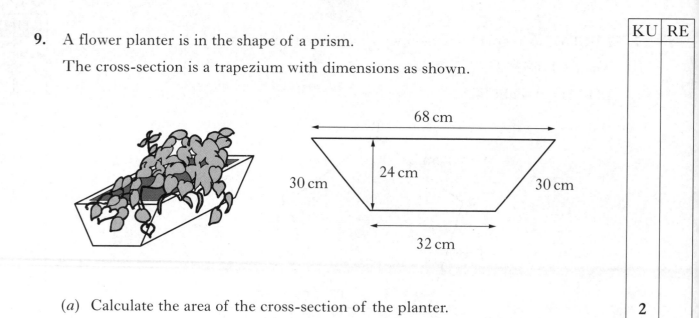

(*a*) Calculate the area of the cross-section of the planter.

2

(*b*) The volume of the planter is 156 litres.

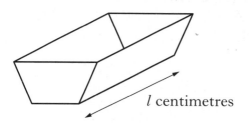

l centimetres

Calculate the length, *l* centimetres, of the planter.

3

10. Tom and Samia are paid the same hourly rate.

Harry is paid $\frac{1}{3}$ more per hour than Tom.

Tom worked 15 hours, Samia worked 8 hours and Harry worked 12 hours. They were paid a total of £429.

How much was Tom paid?

3

KU | RE

11. Paper is wrapped round a cardboard cylinder **exactly** 3 times.

The cylinder is 70 centimetres long.

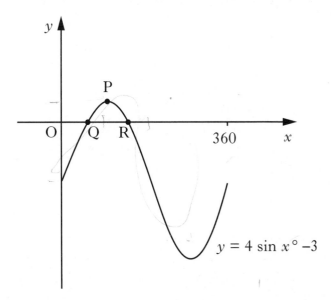

The area of the paper is 3000 square centimetres.

Calculate the diameter of the cylinder.

4

12. Part of the graph of $y = 4 \sin x° - 3$ is shown below.

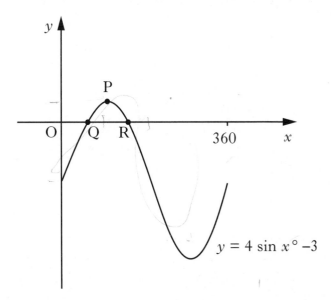

$y = 4 \sin x° - 3$

The graph cuts the x-axis at Q and R.

P is the maximum turning point.

(a) Write down the coordinates of P.

1

(b) Calculate the x-coordinates of Q and R.

4

[Turn over for Question 13 on *Page eight*

KU | RE

13. The diagram shows the path of a flare after it is fired.

The height, h metres above sea level, of the flare is given by

$h = 48 + 8t - t^2$ where t is the number of seconds after firing.

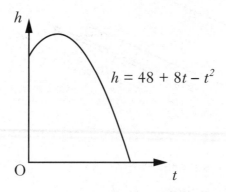

Calculate, **algebraically**, the time taken for the flare to enter the sea.

4

[END OF QUESTION PAPER]

[BLANK PAGE]

C

2500/31/01

NATIONAL
QUALIFICATIONS
2012

WEDNESDAY, 2 MAY
1.30 PM – 2.25 PM

MATHEMATICS
STANDARD GRADE
Credit Level
Paper 1
(Non-calculator)

1 **You may NOT use a calculator.**

2 Answer as many questions as you can.

3 Full credit will be given only where the solution contains appropriate working.

4 Square-ruled paper is provided. If you make use of this, you should write your name on it clearly and put it inside your answer booklet.

FORMULAE LIST

The roots of $ax^2 + bx + c = 0$ are $x = \dfrac{-b \pm \sqrt{(b^2 - 4ac)}}{2a}$

Sine rule: $\dfrac{a}{\sin A} = \dfrac{b}{\sin B} = \dfrac{c}{\sin C}$

Cosine rule: $a^2 = b^2 + c^2 - 2bc \cos A$ or $\cos A = \dfrac{b^2 + c^2 - a^2}{2bc}$

Area of a triangle: Area $= \dfrac{1}{2} ab \sin C$

Standard deviation: $s = \sqrt{\dfrac{\sum (x - \bar{x})^2}{n - 1}} = \sqrt{\dfrac{\sum x^2 - (\sum x)^2 / n}{n - 1}}$, where n is the sample size.

KU | RE

1. Evaluate

$$7{\cdot}2 - 0{\cdot}161 \times 30.$$

2

2. Expand and simplify

$$(3x - 2)(2x^2 + x + 5).$$

3

3. Change the subject of the formula to m.

$$L = \frac{\sqrt{m}}{k}$$

2

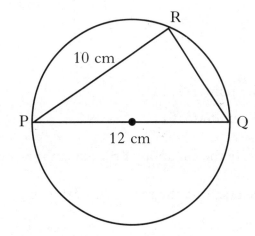

4. In the diagram,

- PQ is the diameter of the circle
- PQ = 12 centimetres
- PR = 10 centimetres.

Calculate the length of QR.

Give your answer as a surd in its simplest form.

4

[Turn over

KU | RE

5. Mike is practising his penalty kicks.

Last week, Mike scored 18 out of 30.

This week, he scored 16 out of 25.

Has his scoring rate improved?

Give a reason for your answer.

3

6. The diagram shows part of the graph of $y = 5 + 4x - x^2$.

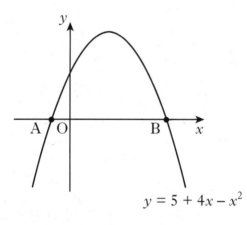

$$y = 5 + 4x - x^2$$

A is the point $(-1, 0)$.

B is the point $(5, 0)$.

(*a*) State the equation of the axis of symmetry of the graph.

2

(*b*) Hence, find the maximum value of $y = 5 + 4x - x^2$.

2

7. Given $2x^2 - 2x - 1 = 0$, show that

$$x = \frac{1 \pm \sqrt{3}}{2}$$

4

8. The graph below shows two straight lines.

- $y = 2x - 3$
- $x + 2y = 14$

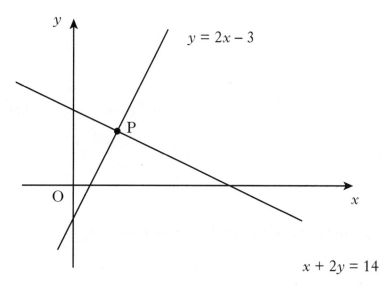

The lines intersect at the point P.

Find, **algebraically**, the coordinates of P.

4

[Turn over for Questions 9, 10 and 11 on *Page six*

	KU	RE

9. Each day, Marissa drives 40 kilometres to work.

(a) On Monday, she drives at a speed of x kilometres per hour.

Find the time taken, in terms of x, for her journey.   **1**

(b) On Tuesday, she drives 5 kilometres per hour **faster**.

Find the time taken, in terms of x, for this journey.   **1**

(c) Hence find an expression, in terms of x, for the difference in times of the two journeys.

Give this expression **in its simplest form**.   **3**

10. (a) Evaluate $(2^3)^2$.   **1**

(b) Hence find n, when $(2^3)^n = \dfrac{1}{64}$.   **1**

11. The sum of consecutive even numbers can be calculated using the following number pattern:

$$2 + 4 + 6 = 3 \times 4 = 12$$
$$2 + 4 + 6 + 8 = 4 \times 5 = 20$$
$$2 + 4 + 6 + 8 + 10 = 5 \times 6 = 30$$

(a) Calculate $2 + 4 + \cdots + 20$.   **1**

(b) Write down an expression for $2 + 4 + \cdots + n$.   **1**

(c) Hence or otherwise calculate $10 + 12 + \cdots + 100$.   **2**

[END OF QUESTION PAPER]

C

2500/31/02

NATIONAL
QUALIFICATIONS
2012

WEDNESDAY, 2 MAY
2.45 PM – 4.05 PM

MATHEMATICS
STANDARD GRADE
Credit Level
Paper 2

1 **You may use a calculator**.

2 Answer as many questions as you can.

3 Full credit will be given only where the solution contains appropriate working.

4 Square-ruled paper is provided. If you make use of this, you should write your name on it clearly and put it inside your answer booklet.

FORMULAE LIST

The roots of $ax^2 + bx + c = 0$ are $x = \dfrac{-b \pm \sqrt{(b^2 - 4ac)}}{2a}$

Sine rule: $\dfrac{a}{\sin A} = \dfrac{b}{\sin B} = \dfrac{c}{\sin C}$

Cosine rule: $a^2 = b^2 + c^2 - 2bc \cos A$ or $\cos A = \dfrac{b^2 + c^2 - a^2}{2bc}$

Area of a triangle: Area $= \frac{1}{2} ab \sin C$

Standard deviation: $s = \sqrt{\dfrac{\sum (x - \bar{x})^2}{n-1}} = \sqrt{\dfrac{\sum x^2 - (\sum x)^2 / n}{n-1}}$, where n is the sample size.

KU | RE

1. There are 2·69 million vehicles in Scotland.

   It is estimated that this number will increase at a rate of 4% each year.

   If this estimate is correct, how many vehicles will there be in 3 years' time?

   Give your answer **correct to 3 significant figures**.

   4

2. Before training, athletes were tested on how many sit-ups they could do in one minute.

   The following information was obtained:

   lower quartile (Q_1)   23

   median (Q_2)   39

   upper quartile (Q_3)   51

   (a) Calculate the semi-interquartile range.

   1

   After training, the athletes were tested again.

   Both sets of data are displayed as boxplots.

   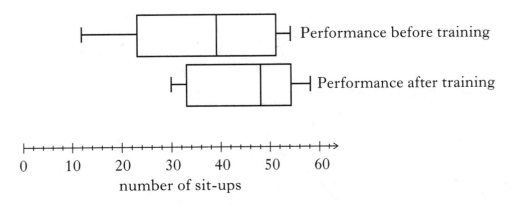

   (b) Make **two** valid statements to compare the performances before and after training.

   2

[Turn over

KU | RE

3. A container for oil is in the shape of a prism.

The width of the container is 9 centimetres.

The uniform cross section of the container consists of a rectangle and a triangle with dimensions as shown.

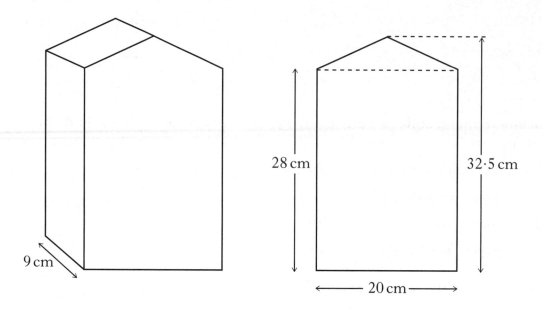

Calculate the volume of the container, **correct to the nearest litre.**

4

4. A sector of a circle, centre O, is shown below.

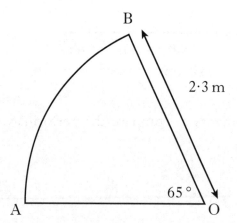

The radius of the circle is 2·3 metres.

Angle AOB is 65°.

Find the length of the arc AB.

3

KU | RE

5. The depth, d, of water in a tank, varies directly as the volume, v, of water in the tank and inversely as the square of the radius, r, of the tank.

When the volume of water is $60\,000\,\text{cm}^3$, the depth of water is $50\,\text{cm}$ and the radius of the tank is $20\,\text{cm}$.

Calculate the depth of the water, when the volume of water is $75\,000\,\text{cm}^3$ and the radius of the tank is $25\,\text{cm}$.

4

6. The price for Paul's summer holiday is £894·40.

The price includes a 4% booking fee.

What is the price of his holiday without the booking fee?

3

7. A heavy metal beam, AB, rests against a vertical wall as shown.

The length of the beam is 8 metres and it makes an angle of $59°$ with the ground.

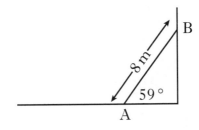

A cable, CB, is fixed to the ground at C and is attached to the top of the beam at B.

The cable makes an angle of $22°$ with the ground.

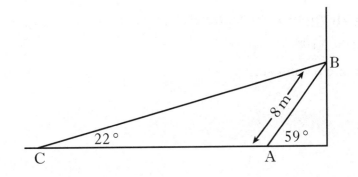

Calculate the length of cable CB.

4

KU | RE

8. A necklace is made of beads which are mathematically similar.

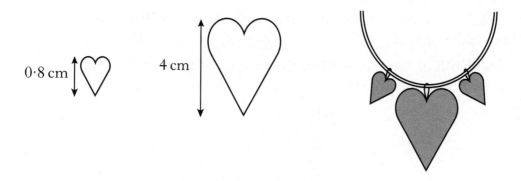

0·8 cm   4 cm

The height of the smaller bead is 0·8 centimetres and its area is 0·6 square centimetres.

The height of the larger bead is 4 centimetres.

Find the area of the larger bead.

3

9. Paving stones are in the shape of a rhombus.

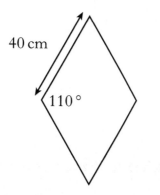

40 cm

110°

The side of each rhombus is 40 centimetres long.

The obtuse angle is 110°.

Find the area of one paving stone.

4

10. A taxi fare consists of a call-out charge of £1·80 **plus** a fixed cost per kilometre.

A journey of 4 kilometres costs £6·60.

The straight line graph shows the fare, f pounds, for a journey of d kilometres.

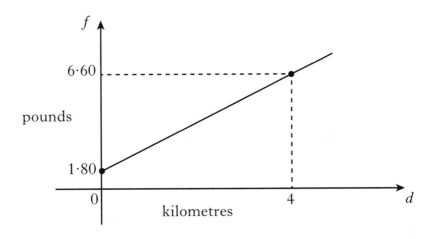

(a) Find the equation of the straight line.

(b) Calculate the fare for a journey of 7 kilometres.

11. Quadrilateral ABCD with angle ABC = 90° is shown below.

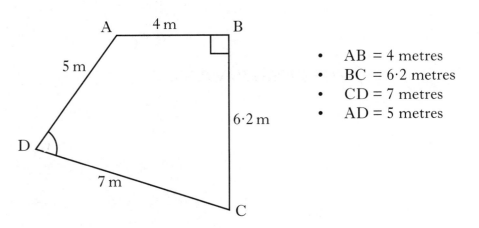

- AB = 4 metres
- BC = 6·2 metres
- CD = 7 metres
- AD = 5 metres

(a) Calculate the length of AC.

(b) Calculate the size of angle ADC.

[Turn over for Questions 12 and 13 on *Page eight*

	KU	RE

12. $f(x) = 3 \sin x°,$ $0 \le x < 360$

(a) Find $f(270)$.

1

(b) $f(t) = 0.6$.

Find the two possible values of t.

4

13. Triangles PQR and STU are mathematically similar.

The scale factor is 3 and PR corresponds to SU.

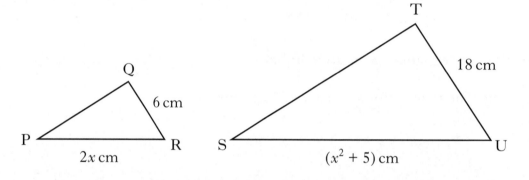

(a) Show that $x^2 - 6x + 5 = 0$.

2

(b) Given QR is the shortest side of triangle PQR, find the value of x.

3

[END OF QUESTION PAPER]

STANDARD GRADE | ANSWER SECTION

SQA STANDARD GRADE CREDIT
MATHEMATICS 2008–2012

MATHEMATICS CREDIT PAPER 1
2008 (NON-CALCULATOR)

1. $5 \cdot 8$

2. $5(x - 3)(x + 3)$

3. $H = \sqrt{\dfrac{W}{B}}$

4. $y = -2x + 18$

5. $\dfrac{3p + 5}{p(p + 5)}$

6. (a) $2(x + 8)$

   (b) $0 \cdot 5x$

   (c) 12 kilometres per hour

7. (a) 5

   (b) $x + 6$

   (c) $7x + 7$

8. (a) $(2, 0), (8, 0)$

   (b) 25 units

9. $m\dfrac{7}{2}$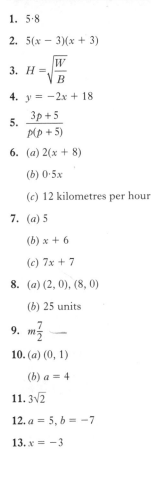

10. (a) $(0, 1)$

    (b) $a = 4$

11. $3\sqrt{2}$

12. $a = 5, b = -7$

13. $x = -3$

MATHEMATICS CREDIT PAPER 2
2008

1. 52 900 tonnes

2. (a) 34, 29

   (b) $\dfrac{11}{30}$

3. £56·25

4. (a) $x + y = 60$

   (b) $50x + 20y = 1740$

   (c) 18 fifty pence coins

5. (a) $\sqrt{65}$

   (b) $\sqrt{40}$

6. No, the boat is not beyond the horizon, with numerical comparison

7. $5 \cdot 62$ m

8. (a) $126 \cdot 9$ m^2

   (b) $90°$

9. (a) $150°$

   (b) $45 \cdot 8$ cm

10. (a) £152·80

    (b) $C = 25d + 0 \cdot 12m - 24$

11. (a) 21

    (b) $55 = \dfrac{1}{2}n(n - 1)$.

       $n^2 - n - 110 = 0$

    (c) 11

12. (a) $78 \cdot 7, 258 \cdot 7$

    (b) $438 \cdot 7$

MATHEMATICS CREDIT PAPER 1 2009 (NON-CALCULATOR)

1. $27 \cdot 11$

2. $2\dfrac{5}{6}$

3. (a) 19

   (b) $t = \pm 7$

4. (a) $(x - 2y)(x + 2y)$

   (b) $2x^2 + 7x - 4$

   (c) $3x^{\frac{3}{2}} + x^{-\frac{3}{2}}$

5. $4\sqrt{3}$

6. P(female) plus justification

7. £200

8. (a) $6x + 2y = 42$

   (b) $5x - 2y = 2$

   (c) $x = 4,\ y = 9$

9. $d = \sqrt{\dfrac{20\,f}{k}}$

10. (a) 14 seconds

    (b) 98 metres

11. $\sin A = \dfrac{6 \sin 30°}{10} = 0 \cdot 3$

MATHEMATICS CREDIT PAPER 2 2009

1. $3 \cdot 1 \times 10^{24}$

2. 8

3. $-1 \cdot 2,\ 5 \cdot 2$

4. $112 \cdot 5$ cm^2

5. £372, £74

6. (a) $y = 2x + 3$

   (b) 43

7. (a) $t = \dfrac{k}{d^2}$

   (b) $8°$ C

8. no, plus justification

9. (a) $20°$

   (b) $55 \cdot 6$ m

   (c) $312°$

10. 10 years

11. (a) 2042 cm^2

    (b) $136 \cdot 1$ cm

12. (a) 210

    (b) $\dfrac{1}{2}(n + 1)(n + 2) = \dfrac{1}{2}(n^2 + 3n + 2)$

    (c) $\dfrac{1}{2}n(n + 1) + \dfrac{1}{2}(n^2 + 3n + 2) = (n + 1)(n + 1)$
    $= (n + 1)^2$

MATHEMATICS CREDIT
PAPER 1
2010 (NON-CALCULATOR)

1. £2.79

2. $\dfrac{4}{11}$

3. $s = \dfrac{2t - 4}{7}$

4. (a) proof

   (b) $x = -1, x = 7$

5. (a) $\dfrac{5}{9}$

   (b) 15

6. 750 grams

7. (a) $2m + c = 7$

   (b) $4m + c = 17$

   (c) $m = 5, c = -3$

   (d) 5

8. (a) 6

   (b) $4\sqrt{2}$

   (c) $\dfrac{3\sqrt{2}}{4}$

9. (a) B$(-6, 0)$

   (b) $x < -6$

10. (a) $\dfrac{5^2 \times 6^2}{4}$

   (b) $\dfrac{n^2 (n + 1)^2}{4}$

   (c) 2025

11. $x = \dfrac{6}{5}$

MATHEMATICS CREDIT
PAPER 2
2010

1. 35 400 tonnes

2. $x^3 - 2x^2 + x$

3. (a) 101, 1·69

   (b) Two valid statements e.g.
   The second sample has, on average, a greater number of pins per box
   The second sample has a greater variability in the number of pins per box

4. $-2·6, 0·9$

5. (a) 0·867 m

   (b) 1·57 m^3

6. 88·0 cm

7. 24 cm

8. 9

9. 5:6

10. 126·9°

11. (a) $h = \dfrac{kV}{b^2}$

   (b) 18 cm

12. $x = 5$

13. (a) 3·875 m

   (b) 3·5 m

MATHEMATICS CREDIT PAPER 1 2011 (NON-CALCULATOR)

1. $2 \cdot 491$

2. $2(m - 3)(m + 3)$

3. -4

4. $-\dfrac{7}{5}$

5. 17

6. (a) 4

   (b) $\dfrac{7}{15}$

7. (a) $2a + 4c = 56$

   (b) $a + 3c = 36$

   (c) £8 and £12

8. (a) $y = -\dfrac{3}{2}x + 12$

   (b) $\left(\dfrac{8}{3}, 8\right)$

9. (a) $2a^{-3}$

   (b) 2

10. $\dfrac{10}{\sin 150°} = \dfrac{4}{\sin B}$

    $10 \sin B = 4 \sin 150°$

    $10 \sin B = 4 \times \dfrac{1}{2}$

    $\sin B = \dfrac{1}{5}$

11. (a) $F \propto \dfrac{s}{d^2}$ or $F = \dfrac{ks}{d^2}$

    (b) reduced to $\dfrac{1}{8}$ of original

12. (a) 55

    (b) $s_n = \dfrac{1}{2}n(n + 1)$

MATHEMATICS CREDIT PAPER 2 2011

1. $40 \cdot 9948$

2. $3x^3 - 14x^2 + 7x + 4$

3. $-2 \cdot 8, 1 \cdot 3$

4. £4500

5. No, as $0 \cdot 879 < 0 \cdot 9$

6. No, as $7754 \cdot 6 \neq 8040$

7. (a) $108°$

   (b) $1 \cdot 62$ cm

8. $10 \cdot 6$ cm

9. (a) 1200 cm^2

   (b) 130 cm

10. £165

11. $4 \cdot 55$ cm

12. (a) $(90, 1)$

    (b) $48 \cdot 6°, 131 \cdot 4°$

13. 12 seconds

MATHEMATICS CREDIT
PAPER 1
2012 (NON-CALCULATOR)

1. $2 \cdot 37$

2. $6x^3 - x^2 + 13x - 10$

3. $m = (kL)^2$

4. $2\sqrt{11}$

5. Yes, plus justification
   For example:
   yes, because $\dfrac{96}{150} > \dfrac{90}{150}$
   or
   yes, because $0 \cdot 64 > 0 \cdot 6$

6. (a) $x = 2$

   (b) 9

7. Proof, for example:
   $$x = \frac{2 \pm 2\sqrt{(-2)^2 - 4(2)(-1)}}{2 \times 2}$$
   $$= \frac{2 \pm \sqrt{12}}{4}$$
   $$= \frac{2 \pm 2\sqrt{3}}{4}$$
   $$= \frac{1 \pm \sqrt{3}}{2}$$

8. $(4,5)$
   Method 1
   - $2y = -x + 14$
   - $4y = -2x + 28$
   - $y = 5$
   - $(4,5)$
   or
   Method 2
   - $x + 2(2x - 3) = 14$
   - $5x - 6 = 14$
   - $x = 4$
   - $(4,5)$

9. (a) $\dfrac{40}{x}$

   (b) $\dfrac{40}{x + 5}$

   (c) $\dfrac{200}{x(x + 5)}$

10. (a) 64

    (b) -2

11. (a) 110

    (b) $\dfrac{n}{2} \times \left(\dfrac{n}{2} + 1\right)$

    (c) 2530

MATHEMATICS CREDIT
PAPER 2
2012

1. $3 \cdot 03$ million

2. (a) 14

   (b) Two valid statements, for example:
   - on average the **number** of sit-ups per athlete has risen
   - the number of sit-ups is less varied

3. 5 litres
   Method 1
   $(28 \times 20) + \left(\dfrac{1}{2} \times 20 \times 4 \cdot 5\right)$ (605)
   $\left[(28 \times 20) + \left(\dfrac{1}{2} \times 20 \times 4 \cdot 5\right)\right] \times 9$
   5445
   5

   Method 2
   $9 \times 20 \times 28$ (5040)
   $9 \times \left[\dfrac{1}{2} \times 20 \times 4 \cdot 5\right]$ (405)
   5445
   5

   Method 3
   $9 \times 20 \times 32 \cdot 5$ (5850)
   $9 \times \left[\dfrac{1}{2} \times 20 \times 4 \cdot 5\right]$ (405)
   5445
   5

4. $2 \cdot 61$m

5. 40 cm

6. £860

7. $18 \cdot 3$ metres

8. 15 cm^2

9. $1503 \cdot 5$ cm^2
   Method
   $\dfrac{1}{2} ab \sin C$
   $\dfrac{1}{2} \times 40 \times 40 \times \sin 110°$
   $751 \cdot 75$
   $1503 \cdot 5$

10. (a) $f = 1 \cdot 2d + 1 \cdot 8$

    (b) £$10 \cdot 20$

11. (a) $7 \cdot 38$ metres

    (b) $73 \cdot 8°$

12. (a) -3

    (b) $11 \cdot 5°, 168 \cdot 5°$

13. (a) Proof
    $\dfrac{2x}{x^2 + 5} = \dfrac{6}{18}$ or $3 \times 2x = x^2 + 5$
    $x^2 - 6x + 5 = 0$

    (b) 5

Hey! I've done it

BrightRED
PUBLISHING

© 2012 SQA/Bright Red Publishing Ltd, All Rights Reserved
Published by Bright Red Publishing Ltd, 6 Stafford Street, Edinburgh, EH3 7AU
Tel: 0131 220 5804, Fax: 0131 220 6710, enquiries: sales@brightredpublishing.co.uk,
www.brightredpublishing.co.uk

Official SQA answers to 978-1-84948-254-7
2008-2012